Molecular Terrorism

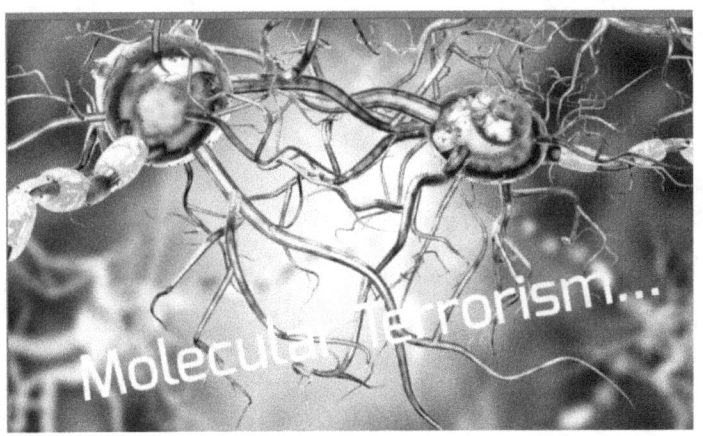

Copyright © 2023 Wayne McRoy

All Rights Reserved

ISBN: 9798866296965

You're listening to The Alchemical Tech Revolution, and I am your host, Wayne McRoy.

Good evening, everyone.

Tonight we're going to discuss molecular terrorism.

What is that, you might ask?

And where did this notion come from?

2:02

Well, tonight we're going to be looking into a book here written by the late Jim Marrs called The Trillion Dollar Conspiracy, How the New World Order Man Made Diseases and Zombie Banks Are Destroying America.

And we're going to read a portion of this tonight, which deals with a subject that perhaps hasn't been touched upon here too much in the past.

2:30

But it is something that's been on my radar for some time now.

And I think it's something we need to talk about.

And this would be a notion within the Medical Sciences called mycoplasmas and prions or prions, I think maybe the appropriate pronunciation for it.

2:51

But it's spelled PRIONS, prions.

So I think we're more familiar with that as a term in and of itself.

So we'll use the term prions, even though it's probably a mispronunciation.

That's OK, I'm going to roll with it.

So never make fun of somebody for mispronouncing a word, because that means that they learned it from reading.

3:14

So if that's the case, remember that.

So we're going to talk about this notion of prions and what exactly are these?

Well, we'll get a little bit into that here and we'll discuss this notion and the possible implications of such a thing.

3:30

And of course, just like everything else that we are handed by the mainstream, I think what we're being handed as far as this goes is also a missed description.

But still there's important aspects of it to be discussed nonetheless.

3:48

So we will do so.

So we're going to begin here.

In the early 1940s, Nazi medical scientists had managed to isolate the bacterial toxin from Brucella bacteria, usually known as brucellosis or undulant fever, and mostly found in mammals, especially cows.

4:09

And they formed it into a crystalline form or agent.

Brucellosis is an ancient bacteria and it was selected because it was insidious, very difficult to detect, and present in almost every organ or organ system of the human body.

4:26

When activated by the crystalline agent, brucellosis stimulates various diseases that prompt a variety of symptoms, including debilitating fatigue, high fever, shivering, aching, drenching, sweats, headache, backache, weakness, and depression.

4:45

Damage to major organs is possible, leading to ailments such as multiple sclerosis, arthritis, and heart disease.

And I'm going to pause for a moment here folks.

Sounds like a veritable who's who of American Society today, doesn't it?

5:03

We have all of these various symptoms, these somewhat debilitating systems, at times symptoms.

Excuse me, but debilitating symptoms at times that are president, people, we have chronic illnesses that hadn't existed in the past, and we have to wonder what's going on with that.

5:22

Perhaps there's something to this notion of prions and its connection back to the Nazi scientists.

Well, let's read on and see what else we can find out here about this this topic, The project paper CLIP.

5:44

Medical scientists coming to America brought with them this toxin known as a mycoplasma, A distinct type of bacteria lacking a cell wall.

A US government report dated January 3rd, 1946 carried a section entitled Production and Isolation for the first time of a crystalline bacterial toxin.

6:07

The Nazi bug had been reduced to a crystalline form, creating an artificial virulent disease agent derived from the original bacteria.

Gonna pause for a moment there, folks.

So we have this record here of the Nazis making this type of crystalline form out of this thing and made it into a weapon.

6:31

Weaponized biologics.

Let's continue reading and see what else we can find out here.

This crystalline bacterial agent could be dispensed by aerial spraying or by infected insects.

6:48

The agent also did not respond to most antibiotics, including penicillin.

Acting as a parasite, it stimulated both bacterial and viral diseases and because it attached to specific cells without killing them, was virtually undetectable by conventional medical diagnosis techniques.

7:09

Such diseases are considered untreatable and usually fatal because they mostly affect the brain or neural tissue.

A couple key points were pointed out in this last paragraph here.

So we see here that it was dispensed through aerial spraying or through infected insects.

7:31

You got to keep that in mind because what is heavily being promoted right now by various organizations throughout the world, well, it's putting insect protein into the food.

Insect protein.

Well, these prions are a type of protein if you want to get down to the nitty gritty of it all, So you got to wonder what's up with that.

7:57

So we see this is what the Nazi scientists initially allegedly came up with, this crystalline version of it that can be sprayed in an aerosol or it could be used to infect insects.

And then the insects, I guess would either sting or bite or some such thing.

8:14

Or perhaps just carry it in them.

And if they get in the food supply, there it is.

And the thing about this is brucellosis.

It didn't respond to antibiotics, including penicillin, which was the big one at the time of the end of the World War here, World War 2.

8:34

So it says here that it acts as a parasite, and that's an interesting crossover as well.

And you'll find if you really begin to delve into the medical science that we have, if you begin to delve heavily into the machinations of how they think medical science and germ theory works, you'll find an interesting correlation here.

9:03

You will find that many of these alleged viruses and bacterial infections, well they act like parasites and some of them respond to anti parasitic medications.

9:20

That's interesting to keep in mind.

So we have here this is a Mycoplasma taken from a bacterium, the Brucella bacterium.

They developed this mycoplasma into a weapon system and it functioned much like a parasite would.

9:44

So it stands to reason to think, if this toxin, which was developed from the bacterium, acts like a parasite, what else could be done with this if it could be weaponized in a way where it can be sprayed in the air or used through a vector like insects?

10:07

A stinging, biting insect or an insect that gets mixed into the food supply and perhaps some of its cell structure winds up in the food supply somehow.

What else can they put this in?

I don't think I have to beat around the Bush too much and saying, is it possible that something like this could end up in some type of a vaccination campaign?

10:34

How much do you trust your government, folks?

That's the question we have to ask.

How much do you trust the big pharmaceutical companies?

How much do you trust these multinational corporations?

And, of course, the nonprofit organizations that decide much of the policy in this world.

10:50

How much do you trust them?

Do you think it's out of the realm of possibility they would do something like this?

As much as I would like to give these people the benefit of the doubt, they seem to prove time and time again, they're not trustworthy, don't they?

11:07

They lie to us ad nauseam about everything.

They knowingly 'cause harm to a great number of people.

I'll remind you, Pfizer was involved in the biggest lawsuit ever levied against an organization, and they lost and had to pay out billions of dollars in the settlement.

11:31

But they would rather do that than disclose what they're really doing with their products because it's cheaper for them to do so to risk the legal action being taken against them.

So we have to keep these things in mind, but that's an interesting little aside.

11:53

So it also says here such diseases are considered untreatable and are usually fatal because they mostly affect the brain or neural tissue.

And oftentimes this is true of parasites as well.

At some point, given widespread toxicity from these things to the body, they do go into the neurological system and often times they'll set up shop there.

12:20

And that's usually when you're near the end stages of these types of things.

This is documented with parasites as well as various other forms of illnesses, and they describe many of these things as viruses.

12:37

And I know some people take umbrage with the notion of the existence of a such a thing as a virus.

You could call it whatever you want.

I mean, logically, what it really is, is it's little bits of protein.

What a virus is, is it's a little, little bits of protein.

12:54

It's it's matter that doesn't really belong within the confines of a cell.

It's extracellular matter that exists and can affect the cell structures in certain ways.

So if you want to get technical with all of that we have, this kind of thing does potentially exist, and people will oftentimes argue over the validity of germ theory versus terrain theory and this kind of thing.

13:23

I think the truth lies somewhere in the middle.

I don't completely discount the tenets of germ theory, because, let's face it, sometimes it works.

Sometimes it works.

How many people have not taken an antibiotic to treat an illness and improved because they've taken this antibiotic?

13:43

Sometimes it's necessary.

And that's the whole thing we need to recognize here.

Although we should not solely rely on this notion of allopathic medicine for everything, sometimes it is necessary and sometimes it does work and it saves lives, And I think we can all acknowledge that.

14:08

So even though our modern medical system has its flaws and many of the foundations of it are faulty, we do have oftentimes techniques that work.

And this is important to keep in mind because oftentimes they do work and there's a time and a place for allopathic medicine.

14:26

And I think sometimes it is necessary.

But there are other forms of medicine that have often gotten overlooked in the modern era.

And this is where the whole germ theory versus terrain theory argument sets in.

Because many a doctor has realized, you know what, this is not being adequately explained to me how this works.

14:47

It seems something's amiss here.

And they began to put two and two together and found the research going back looking at the work of Louis Pasteur and of course the others of that time period, Beauchamp.

15:08

And they found that the terrain theory seems to have a lot more provable types of associations with it than much of germ theory does.

But like I said, I will digress on that point, because I think the truth lies somewhere in the middle.

15:30

I don't think terrain theory is absolutely 100% spot on, and I don't think germ theory is absolutely 100% spot on.

I think the truth of how things operate may lie somewhere in the middle.

And maybe we have some good notions, good ways of looking at things, some good methodologies from each of these different systems or ways of looking at it, and they can produce results in certain instances.

15:55

But we can't lose sight of these things that we do have that are documented within the auspices of our modern science.

And a lot of this stuff, of course a lot of this stuff, of course, gets heavily utilized by the mainstream.

16:17

They like to use this notion of viruses and all kinds of such things like that as a descriptor for many of these diseases and illnesses that they lack understanding of.

And of course they stick very wholeheartedly to the germ theory because that ties very heavily to the allopathic medicine system.

16:39

So that being the case, we have some of these commonalities that go on.

So whatever your stance is on the topic, understand when we say virus you could think of it as just little bits of protein free floating protein that are not part of a cell structure.

So is it waste material maybe or is it something else?

16:58

Is it a type of infection vector in and of itself, Possibly.

Or could it be a both?

It could be a combination of both of those things.

Maybe it is cellular waste products, but if you have enough garbage floating around that in and of itself can cause illness or dis-ease in a system.

17:20

So think about it that way.

So even if it is just, say, garbage expelled from cells, if there's a lot of it and the cells are swimming in it, that's like swimming in a septic tank for a human being.

17:37

You wouldn't want to do that, would you?

Well, same thing with your cells.

So whatever your notion of this descriptor, of what they call the virus is, or all of these associations made with it, we could set that aside and agree that there are these little bits of protein, or whatever it is that this stuff is made out of floating around.

18:03

And perhaps that has something to do with this notion of prions, as we'll see as we get a little further on.

But notice that this was derived from a bacterium from bacteria.

Bacteria is a well documented thing.

It most certainly does exist and we could see it under microscopes and observe certain facets of how it works, cell division and things like that.

18:27

So we we know that there are indeed bacteria.

So we have this vector has been taken from bacteria and it's been weaponized was weaponized by the Nazis and brought over here by project paper clip scientists from Nazi Germany.

18:43

And I'm of course it probably went to Russia as well and various of the other nations that picked up Nazi scientists because America is not the only one but our project paper clip is probably the best known of that.

So that's why it's it's named here but let's continue on.

19:02

I've rambled on long enough about that.

These sub viral bacterium particles have various names.

They have been termed prions by Nobel Prize winner Doctor Stanley B, Prisoner Stealth Viruses by Doctor John Martin of the Center for Complex Infectious Diseases, Amyloids by the late Doctor Carlton Gadusic, winner of the 1976 Nobel Prize in Medicine for his work on mysterious epidemics at the National Institutes of Health.

19:34

And of course, Mycoplasma brucellosis by Donald Scott and Garth Nicholson.

So I'm going to pause for a moment here, folks.

So these little bits of fragmented pieces from this bacterium that they were able to make into a crystalline substance were named or termed prions by Doctor Stanley prisoner, Nobel Prize winner, Doctor Stanley prisoner.

20:04

And of course, these things have been called many different things.

And do they equate to viruses?

Well, it says here that these are sub viral bacterium particles.

So these bacterium particles are smaller than alleged viruses.

20:20

So if you have a combination of these various materials, could they perhaps constitute what they would call a virus?

Maybe.

Here's the the the nuts and bolts of all of it.

OK, we have these little particles that are smaller than bacteria.

20:38

They're smaller than cells.

They can interact with cells, and they call them all different things based upon their size and their shape and how they function.

And sometimes they do function in ways in which they interact with the cells and cause the programming of cell death within the cell.

20:56

And sometimes they just get absorbed by the cell and excreted.

And we have different names for these.

So we have these tiny particles that are called prions or stealth viruses, amyloids, various other names for them.

21:16

And that being the case, we can acknowledge the birth of this within this Nazi program.

So this is where they first began studying this and noticing this stuff and using this stuff.

21:34

So what's the true nature of it?

It's hard to say, but let's continue on and see what else we could get out of this.

According to a paper by Stanley Prisoner prions are unprecedented infectious pathogens that cause fatal neurodegenerative diseases by the entirety, entirely novel mechanism of altering proteins in the body.

21:57

Prion diseases may present as genetic, infectious or sporadic disorders, all of which involve modification of the prion protein, wrote Prisoner.

Going to pause for a moment here, folks.

So these prions are sub particles that combine with the proteins or make up certain proteins.

22:17

And a lot of this, they say, is about how the proteins fold the folding of the proteins.

Now I know this may sound a bit dry to some people out there.

We'll get to the good stuff here soon.

Just trying to lay down the basic foundation of what is said in the accepted science here so we can move forward.

22:39

Paper clip scientists working on these infectious organisms were based primarily in laboratories at Fort Detrick, MD, Cold Spring Harbor, NY and Edgewood Arsenal, Maryland.

It was here and in hundreds of other laboratories throughout America that immediately after World War 2, our former enemy scientists were brought in under Operation Paper Clip to continue their research and development of some of the most horrible weapons of mass destruction known to mankind, noted molecular researchers Garth and Nancy Nicholson.

23:12

In their 2005 book Project Day Lily, The husband and wife molecular researchers noted there are 200 species of Mycoplasma.

Most are innocuous and do no harm.

Only four or five are pathenogenic Mycoplasma fermentans, which is Incognitis strain probably comes from the nucleus of the Brucella bacterium.

23:38

This disease agent is not a bacterium and not a virus.

It is a mutated form of the Brucella bacterium combined with a Vista virus from which the mycoplasma is extracted, they said.

The little mycoplasma also lost some of its genetic information, such as the genes that encode the thick cell wall and other genes that code for certain enzymes and metabolic pathways.

24:04

Thus, it is smaller than the most common bacteria, and without the distinctive cell walls found in most bacteria, it can take on a variety of morphologies.

Going to pause for a moment here, folks.

So this thing can take on many shapes.

24:22

Many different shapes and combinations.

Let's read on.

It must hide inside animal or human cells to survive.

And although originally thought to be fairly fragile, though, little mycoplasma was hardier than anyone had ever imagined.

24:41

Gonna pause for a moment here, folks.

So the Nazis had actually developed a very effective bio weapon with this.

They caused it to mutate in a way where it's turned into this type of an infection vector.

25:01

Let's continue on.

Although considered primitive by bacteriological standards, the Mycoplasma actually evolved from bacteria that contained cell walls but lost its ability to make its own cell wall, probably because it no longer needed it when hiding inside host cells and tissues.

25:20

Going to pause for a moment here folks, and this year I would call total poppycock because they always have to push that in there.

I don't accept the theory of Darwinian evolution as presented.

I don't think these things happen that way.

25:38

It seems to me this was probably a deliberate mutation made by the Nazis, The Nazi scientists who were working on this, when they were able to form it into a crystalline type of a structure and weaponize it as such, well, it continued to reproduce without the cell walls and stuff like that, all that tissue.

26:04

So they admit it was a type of a mutation from this strain of Brucella bacteria that they used and somehow it was modified and now all of a sudden they had this really effective type of a weapon system, this mycoplasma or prion if you want to call it that.

26:22

It's a sub viral or smaller than viral particles.

So let's read on.

So it says.

But it made-up for the loss of some of its genetic information by having evolved with other genetic sequences that allowed it to enter and colonize cells.

26:39

Just like a virus is going to pause for a moment here, folks.

It evolved, of course, all naturally, right?

That's how this all happens.

It just so happened to evolve in a Nazi lab into this, into this morphology of sorts, just like viruses.

26:58

Amazing how they do that, isn't it?

How they just randomly will evolve into something that can do all of this harmful stuff to creatures.

Even though it doesn't make sense if you look at it from the Darwinian evolution standpoint, you see, if you believe what Darwinian evolution tells you, well, it's the survival of the fittest.

27:21

And a creature and Organism would want to somehow or another maintain its life or prolong its life.

So when you have something like this that evolves into something so pathogenic to a creature that it can no longer perpetuate itself doesn't make sense.

27:42

It can't perpetuate its species.

If it's infecting a creature and it kills that creature so quickly, then it cannot continue the life cycle thereof.

So that doesn't make sense if you want to look at it from the evolutionary biology standpoint.

28:01

But of course they make other types of claims with this.

They throw common sense out the window.

In modern science, with a lot of these things, you dare not question their sacred cow of Darwinian evolution.

Of course it makes sense, and I'm sure there's probably some science minded individuals out there would who would probably explain it to me because I I just can't see it.

28:25

Because, you know, maybe I'm not educated enough to understand the tenets of microbiology.

So, you know, more power to you, but I'm not buying it.

But at any rate, let's go ahead and continue on.

So, but it was not a virus because it retained the genetic and biochemical remnants of bacteria.

28:48

Like a virus.

However, it damaged cells by interfering with some of the cell's biochemical cycles and it encoded some nasty molecules that caused invaded cells to slowly self destruct and die, said the Nicholson's, noting that important targets inside cells were the mitochondria, the cellular batteries that produce energy and the DNA.

29:11

Going to pause for a moment here folks.

The mitochondria is the powerhouse of the cell.

If you took any kind of college level biology or anything like I did, I remember this.

Of course I'm showing my age because I'll tell you, I bet you it's changed an awful lot since then.

29:29

I have no doubt I'm showing my age.

I mean, I'm talking this is going back to the early 1990s when I was taking these courses and I remember a little bit of this stuff and things have changed quite a bit in those thirty some odd years.

29:47

I'm sure they claim to know a whole lot more now than what they did then, but that's neither here nor there.

So the mitochondria, the powerhouse of the cell and the DNA, so that were those were the important structures within the cells that these prions affected.

30:04

The Nicholsons explained that biological warfare research conducted between 1942 and now has created more deadly and infectious forms of mycoplasma.

Continuing the work of Nazi scientists, researchers in the United States weaponized the mycoplasma by reducing the pathogen to a synthesized crystalline form.

30:25

They later tested it on an unsuspecting public in North America.

Going to pause for a moment here, folks.

No gasp.

Never.

They would never do that, right?

How many people would tell you they would never do that?

They most certainly have done that in the past, and I've no doubt that they continue to do such things today.

30:47

They do not think like you and I when you're thinking in your mind or saying out loud, they would never do that.

What you're saying is I would never do that.

Just remember, because it's something you would never do to another human being, doesn't mean that these psychopaths who run things wouldn't.

31:04

They most certainly would, as they've proven time and again.

Like I said, I would really love to give these people the benefit of the doubt and think that they're the good guys working on the good side of humanity, but time and time again they disappoint and they prove me wrong in that optimistic outlook.

31:22

So you have to really take it all in with a grain of salt and understand that you're probably being lied to.

You're probably being experimented on.

You're probably being victimized by some of these various quasi governmental organizations and corporate interests around the world working towards the same goals here with their think tank groups and everything else weaponizing stuff like this.

31:54

Who knows what they've unleashed on us folks.

I mean, look around and what's been going on here the past several years.

Look at how things have exponentially just gotten worse and worse, especially in the healthcare field.

32:13

And wonder.

Look at it all with wonder and understand perhaps there's something more here they're not telling us.

Let's continue on here, though.

So, according to the Nicholson's, the US military's fascination with building this kind of biological weapon lies in the fact that the creature will hide inside cells and cause unbelievable havoc.

32:38

It will destroy the mitochondria, eventually sending cells into an unrelenting death program, and in the process, gene expression will go crazy and surrounding cells will become damaged.

This bug will then escape from its dying host cell and go to other places to eventually colonize every organ.

32:58

And because pieces of the cellular membrane are dislodged when this little mycoplasma leaves its cellular hiding places, its victims should also be presented with an array of autoimmune symptoms similar to those found in various degenerative illnesses.

33:16

It may even mimic some neurodegenerative diseases.

It's beautiful because it should cause diseases such as multiple sclerosis and rheumatoid arthritis, but no one will ever guess that they are caused by an infection.

Most physicians will never figure this out.

33:34

What a delightful weapon.

Several researchers, including the Nicholsons Doctor Leonard G Horowitz, Doctor Joseph S Puleo, and authors of the Brucellosis Triangle, Donald West and William LZ Scott, have linked this Mycoplasma pathogen to a host of increasingly common neurosystematic diseases such as Alzheimer's, bipolar disorder, Crohn's colitis, chronic fatigue syndrome, Creutzfeldt-Jakob,diabetes, dystonia, fibromyalgia, Huntington's, lupus, Lyme disease, multiple sclerosis, myalgic encephalomyelitis.

34:20

Going to pause for a moment here folks.

There's more.

Stay tuned.

Parkinson's disease and even schizophrenia.

And I'm going to pause and I'm going to add my own little addendum and speculate possibly autism as well, because you'll find autism and schizophrenia have a lot in common.

34:50

And in fact autism most of the time prior to being identified in the modern era as such, was often times going even right up into the 1960s and 70s, anybody presenting with autism symptoms, they were diagnosed as a pre schizophrenic.

35:10

Any child who was presenting autism symptoms, they were diagnosed as being pre schizophrenic.

They had this intrinsic connection in this way.

So I would argue possibly another symptom or another illness or dis ease being caused by these mycoplasmas.

35:36

Perhaps.

Not saying that's definitely what's going on, but you have to wonder when you have people putting out this kind of information and you go back and you trace the roots of it and find, well, it just kind of popped up toward the end of the World War Two period within the Nazi sciences came over here to America.

36:02

And look at the chronic illnesses that permeate our culture today and think about this is the origins most likely of where this stuff came from.

I don't accept for a second that it existed in the past, but we just never recognized that or knew what it was until the Nazis isolated it and decided to weaponize it.

36:28

I don't accept that.

Maybe there's some modicum of truth to the notion that this waste material given off by cells or these particles, maybe they existed in some form, but not in the very potent and very toxic type form that these things have been weaponized and mutated purposely into.

36:53

And this is when we see the onset of a lot of new neurological disorders, new immunological disorders, autoimmune disorders, correlated to the beginnings within this time period.

37:09

If you go back prior to the 1960s, much of this stuff did not exist.

So you have the paper clip scientists bringing over the Nazi work on this.

And of course, we see the rise in many of these autoimmune disorders and all of these different things that were just described in this last paragraph here, these mycoplasmas or prions being connected to the Nazi bio weapon.

37:40

Let's read on here though, because it's not over even there with schizophrenia.

It says here some strains of Mycoplasma are now being blamed for cancer and AIDS.

According to the former chief virologist for the pharmaceutical company Merck, the late doctor Maurice Hillman, this disease agent is now carried by everybody in North America and possibly most people throughout the world.

38:06

Going to pause for a moment here, folks.

If you've never seen Doctor Maurice Hillman talking about what's known as SV40, Simeon Virus 40 and how it found its way into the polio vaccine campaign, you need to look up that video and watch it.

38:28

He just talked very openly about it.

They knew it was there.

They knew that it caused cancer and possibly it may have possibly been connected to the AIDS epidemic in some such way.

Now there's a lot of lies and deception that attached to the AIDS epidemic, so I'm not sure what to believe there with that.

38:50

But most certainly SV 40 has been a known vector for causing cancer in human beings and it was present in the polio vaccines and Merck Pharmaceutical knew about it and did nothing to remedy the situation.

39:10

You see it permeated all the batches and they didn't want to have to lose all the money by doing a massive recall and of course trying to reformulate things and get this contaminant out of there.

That's what they called it, a contaminant.

So the baby boomers around the world who got this polio vaccine are now carrying a cancer causing agent in their cells, stealthily waiting there.

39:43

And many of these things, they're not rapid onset, they take a long time to develop.

So we have this cancer causing agent out there and of course this is all admitted.

This is the stuff that's admitted, OK, openly, publicly.

40:02

There's probably more that's gone on behind the scenes that.

Will never get disclosed, disclosed to the public with this stuff, but certainly we see this.

So I'm assuming this Mycoplasma, these prions had something to do with this SV40 contaminant according to what's being presented here.

40:20

But let's go ahead and we'll read on Mycoplasma.

Researchers claim many people today suffering from various neurological diseases are actually I'll with brucellosis.

Going to pause for a moment here, folks.

So if you have one of these said diseases, some people who've studied these things say that it's quite possible that you're suffering from Brucellosis was a known vector, a known disease back in the times of World War 2, largely been forgotten about.

40:57

And because they've taken it and weaponized it, separated it from the bacterium portion of it, it acts in a stealth way.

It sits dormant in your body, in your cells, until such time that some external toxin or some internal state of being triggers a mechanism in it, and it activates an activation type mechanism.

41:25

And that could be any number of things.

But when it does, then you have the onset of various of these symptoms, especially the ones relating to these neurological diseases.

You know, multiple sclerosis, Huntington's lupus, Lyme disease, myalgic encephalomyelitis, Parkinson's schizophrenia, probably autism, chronic fatigue syndrome.

41:53

So that's what some of these people claim, who've studied these mycoplasmas, these prions.

So let's continue on here.

That's what their speculation is.

Of course, you'll never see any scientific studies done to try to prove nor disprove this.

42:11

It's not something that the mainstream science culture wants done or the Pharmaceutical industry because, well, it would cause an enormous amount of problems because then there'd be a lot of finger pointing who who let this loose on the world.

42:31

Why are we just continuing to treat symptoms for this?

And how did this happen?

What do we go?

Where do we go from here?

How would we develop some type of a cure or a treatment for this?

42:50

And that would open doors they don't want to open.

Let's face it.

Anyway, let's go ahead and continue though so it says here.

However, because the disease toxin pathogen has been isolated from the source bacterium in a crystalline form, there is no blood or tissue test that will confirm this fact.

43:08

Going to pause for a moment here folks, and once again, no way to prove that it's there.

That's the other problem, of course.

We have that wonderful new tool called PCR.

Maybe they could find it that way, right?

43:27

What do you know?

Let's read on.

Weaponized mycoplasmas generate ammonias that are deposited into the infected cell nuclei.

These nasty beasts intertwine with the genetic machinery and are intracellular rather than intercellular.

43:45

Other infectious agents are involved in the afflicted individual.

These agents are usually mosaics of naturally occurring bacteria and viruses, and the effect upon the afflicted individual depends upon the individual's genetic predisposition and immunological makeup, stated Garth Nicholson.

44:04

Each person is affected differently by the infection, but all afflicted individuals share a constellation of symptoms.

Going to pause for a moment here, folks.

So could it be possible that we have just this one weaponized vector causing massive autoimmune disorders and illnesses of various types in people that they're treating as all these many, many different things?

44:33

Using different medications specifically geared towards these different conditions, these different medical conditions, and treating people for something that may not really be the underlying cause of what's going on?

44:50

You think it's possible?

I think it's most certainly possible, but let's see here.

Because, he says not all these individuals share the same symptoms, but there are a constellation of symptoms, so let's read on, he says.

45:08

We have a survey that describes 120 signs and symptoms.

In the case of the pathogenic mycoplasmas that we investigated, we found the HIV one envelope gene associated with the Mycoplasma.

This gene renders the Mycoplasma more deadly.

45:27

I have always wondered how many people that may have diagnosed as HIV positive actually have the chimeric A mosaic of the mycoplasma bacteria and HIV.

Gonna pause for a moment here folks.

Now, if you go back to the beginnings of this COVID nonsense and the rollout of this COVID vaccine, what were they saying they were saying about AIDS having a part with this?

46:02

Is it beginning to add up?

Are the dots connecting for you yet in the vaccine?

Now we also have testimony coming forward that Pfizer knowingly put forward this vaccine and did not disclose that there were SV 40 particles in it, the very same things that are known to be cancer causing agents that were in the polio vaccine that's come forward.

46:35

So we have this connection.

We have the SV40 factor.

We also have the AIDS factor.

Of course, SV40 was largely connected to the AIDS epidemic as well.

We have AIDS tied to this associated with it.

Are these all separate things going on?

46:53

Is it just a coincidence that all these different vectors have these commonalities?

Is this thread something that's really there?

I'm reading from a book here written by Jim Mars folks, The Trillion Dollar Conspiracy.

47:14

The Trillion Dollar Conspiracy by Jim Mars.

Jim Mars died a few years ago.

I'm going to go ahead and look and see what's the publication date of this book.

This was a reprint edition July 5th, 2011.

47:31

How prescient was Jim Marrs folks?

2011?

This book was written, published, I should say written before then.

And here we are.

Jim Mars died few years back.

47:48

I think it was prior to all the COVID nonsense.

But once again we see he was looking at this stuff and wouldn't you know there's connections all here cropping up today in current events.

Let's read on here.

48:06

So it says.

Reportedly there are 10 strains of HIV.

HIV1 promotes AIDS by compromising the immunization system whereas HIV 2 does not promote AIDS.

The other eight HIV strains are included in the bio warfare arsenal.

48:25

The pathogenic Mycoplasma can promote a non HIV AIDS that mimics the symptoms of AIDS.

No one will talk about this, said Nancy Nicholson.

The mycoplasmas have been genetically engineered with pieces of genetic material from other pathogens such as Brucella.

48:45

The micro plasmas are often cofactors with the Lyme disease microorganism.

All these emerging diseases correlate to a bio warfare experiments conducted during the Cold War that went seriously awry.

Remember, the US did approximately 208 open air tests on the US population without their knowledge or consent over a 30 year period.

49:09

Going to pause for a moment here folks.

So could what we see really accurately be described as a bio weapon?

Quite possibly.

Quite possibly.

49:27

Look at the connections.

I'll let you connect the dots.

Remember back, if you've been looking at this stuff, scrutinizing it, asking proper questions, you'll see all of these connections.

Maybe there's something to this, maybe there's not.

49:44

Or maybe this is one of those notions where they send us down this rabbit hole that never truly seems to resolve and we're grasping after wind and chasing it down.

50:03

Maybe that's the case.

I don't know.

I don't have the scientific chops to really pursue these types of avenues of thought towards this.

Nor do I have the absolute steadfast belief in the way modern science describes these things to us, to trace it down in this kind of a manner.

50:26

But like I said, how prescient was Jim Marrs in writing this back in 2011?

It was published and we see the events of the past couple years and we see how all these things are front and center again.

They have all these commonalities, all these connecting points, and look at the state of society today.

50:47

Look at the died suddenly phenomenon that's been going on in association with this and they deny, deny, deny, deny.

And they continue to tell you of course it's safe and effective.

Go get your new booster shot to protect you from the new variant.

51:06

There's always a new variant.

They want us fearful folks.

They want us to line up and take the nice eugenics based bio weapon, the depopulation weapon, and they're going to do their best, make sure they get what they want.

51:27

They've tried and in my view they've failed to a degree with this program that they've rolled out on the world.

Even though they got a massive amount of the population to buy into the nonsense, there's still enough of us out there that haven't bought into it.

51:45

And now there's people waking up in droves now that they've already already been used as an infection vector Now, but they're waking up to this and they're realizing what's been done, something doesn't add up.

52:01

And that's good news, because if even the ones that were hardcore for this type of a reaction in society are beginning to understand we were lied to, there's bad things associated with this that are coming to light now.

If they're realizing it, then the people rolling out these programs on the human population are going to be in big trouble very soon.

52:27

So they're doubling down their efforts.

They want people in fear.

They don't want people to look at this.

Let's put it that way.

There's something about this whole COVID campaign, and I'll call it a campaign because I think that's what it really is to campaign.

52:44

It was a massive SIOP that may have had some real repercussions to it, but largely it was a campaign.

So this COVID campaign, people are now beginning to get sick and tired of the continuing campaign of the fear being pushed at them from a an invisible enemy.

53:10

And of course, they always have to have some type of an invisible enemy that you can't see coming, that you need to be fearful of and lock yourself in your home.

Be afraid.

Go lock yourself in your home.

Separate yourself from everybody else.

53:26

Stay locked down.

Are you beginning to notice a pattern?

Lockdowns?

Well, going to talk about something here not directly related to this topic tonight.

53:41

We're talking about with this molecular terrorism or these bits of prion material being associated with these various things and possibly as a weapon system turning up in places where people don't suspect.

53:59

But I'm going to talk about another factor here.

So we have, we have many of these people waking up to the medical SIOP that's been played on them now.

They're not as afraid of this COVID campaign anymore.

54:15

They're not lining up in droves like they once were for the latest booster shot.

And they're not in large numbers, masking up, distancing themselves 6 feet apart, wearing gloves, locking themselves in their homes.

54:34

So what's happened over the course of the past couple weeks here on the world stage?

Well, they have new fear vectors for us to be afraid of.

They have two massive proxy wars going on and all the fear and nonsense that associates with them.

54:52

And of course, in conjunction with that, of course they're trying to drum up this whole notion of escalating hatred towards certain racial groups, Let's put it that way.

55:08

You may have noticed if you actually will turn on the television, which if you don't, I don't blame you.

But if you turn on the television, you will have noticed within within just three days of this whole onset of this action, this military action in Israel, in the things going on there with Hamas and Israel from the inception of that.

55:33

When that first rolled out, within a factor of three days, I started to begin to notice on the television.

There were commercials on the television speaking about the dangers of hate speech directed towards a certain people group.

55:54

And I'm not going to mention that certain people group, because you know what folks?

There are certain things even on free speech platforms, certain things you can't really say out loud.

56:11

But I think you know the people group I'm talking about.

There are certain sacred cows, certain things that must not be questioned openly or directly.

And if you do well, you get all kinds of backlash and your stuff begins to disappear from the Internet.

56:35

I don't want that to happen.

So I'm not going to directly name this, but watch and see.

There are certain groups associated with this certain racial group or people group here in America associated with the people over there in that conflict.

56:52

And they have these, this campaign, new campaign talking about how this hatred is a growing threat and it needs to be dealt with.

And if you suspect any such thing, you need to do something about that and report that and it's not acceptable.

57:14

And we need to be against violence towards that, this or you know, hate towards this group.

And don't get me wrong, I don't hate any people group and I don't promote violence towards any group or any such thing.

I just found it interesting that all of a sudden, seemingly out of the clear blue sky, within three days of this launching of this world stage event, we have these commercials cropping up all across the television channels and networks.

57:44

And I've seen them a few Times Now.

And it's gearing up a mindset in the Western culture, public view where they're trying to escalate the idea of this tension between people.

58:06

And of course you see protests going on on the college campuses and every such thing.

If you pay attention to what they're pushing in the mainstream news, well, of course you have these protests, and sometimes they seem to turn violent.

And these certain people groups are resentful of each other and they're fearful of violence against one another.

58:28

And especially on college campuses where they've claimed to be locked in dorms.

Well, what's involved here?

Lockdowns.

They willfully lock themselves in.

So we see the lockdown narrative once again.

58:46

Now Fast forward to was it yesterday?

And we have another alleged mass shooting event in the state of Maine.

And according to news reports, people from as far as 50 miles away.

59:02

We're locking down their businesses, locking down their homes willingly, willfully locking themselves in the lock in.

Again, the lockdown notion coming back around.

You see, it seems to me that these dark occultists who run things and these social engineers at the top of the power structure that want to have their way with the public, they've decided they're going to shift gears and find a different way to lock you down, to lock you in.

59:36

People just aren't buying into the pandemic notion anymore, and they're not going to allow themselves to be locked down for those reasons.

Again, they realize this, so they're shifting gears.

So they're going to go ahead and conflate violence.

59:56

They're going to conflate this notion that there's this mass epidemic of violence occurring, hatred and violence, hate crimes, violent uprising, lone nut individuals going out on rampages for various reasons.

1:00:17

And I can tell you that for the past several months now in the area where I live.

There have been in the news and in the community a number of Times Now where somebody was making threats against local school districts.

1:00:39

The schools have been closing down and locking down, locking down, keeping the kids at home, send the kids home, lock down the schools.

I don't know if this is happening across the country, but if anybody has feedback for me on that would be interesting to see because this is local to my area that this has been happening now several weeks back, going back probably oh maybe a month or so.

1:01:08

They claimed to have caught somebody who was making some of these threats they allegedly worse from somewhere in Texas.

Why would they be calling and making threats to a school district in Pennsylvania, numerous, multiple school districts in Pennsylvania?

But this is the notion.

They wanted us to believe with that.

1:01:23

And now, lo and behold, it's happening again.

What's the reasoning behind this?

They want us to lock ourselves in, lock ourselves down with consent, with our own consent to do that.

So they're going to approach it in a different way.

1:01:42

That's what I see going on as far as that goes.

And like I said, that does kind of divulge and diverge a little bit from the subject and topic here tonight.

But I think these dark occultists and these social engineers of this world that are looking for certain things, I think they pretty much got a good portion of what they wanted with this bio warfare campaign and now they're focusing on subjective biological warfare.

1:02:11

That's what this is when you're looking at trying to coerce people into locking themselves in their own homes and be fearful of another invisible boogeyman.

Some such thing as, I don't know, a lone nut out there, a lone nut predator out there doing stuff, or some hate group out there disseminating hate towards other people just based upon whatever profile they have, whatever racial or religious profile or any such thing.

1:02:46

So they want people to be fearful and willingly lock themselves in and stay out of these public places, close down businesses based upon this or these acts of violence, shutting down school districts and perhaps businesses having people lock themselves in their homes.

1:03:06

Look at, like I said, the example of this alleged event that happened in Maine.

And we can very much see the shifting of the gears in that direction if you're paying attention.

I pay attention, folks.

1:03:24

Don't get me wrong, I don't like watching television, and I watch very little of it, but I do stay tuned in to these different social factors.

That's the only reason I watch television anymore is to rip it apart and understand what kind of programming motif they're using with this stuff.

1:03:44

So that is another subjective biological warfare tool, The media.

Let's put it that way.

But anyway, let's get back on topic here, because I rambled a little far off course.

But that is an important bit of information that perhaps you should start noticing and maybe speaking up about.

1:04:06

So keep that in mind.

Look at it.

Understand this whole trope, the archetype of the lockdown narrative is there again, they're just approaching it a little bit differently this time because they understand the same threat is not going to work for people at some point.

1:04:26

Honestly, I would expect that they're going to pull the alien card, the whole Project Blue Beam idea, and maybe have people walk themselves in their homes willingly because of the alien invasion or some such thing.

We'll see how it goes.

We'll see.

It depends.

1:04:42

How many, how many people can they get to buy into this notion of fear.

The fear of World War escalating combined with the fear of radical Lone wolf gunmen out there going crazy on rampages.

1:04:58

And of course all of the racial divide that they've been gearing up for a long time.

And all of the different profiling of various groups and specific groups that they don't want mentioned or touched, ones you can't speak out against in any way, shape or form.

1:05:18

You get my drift.

But let's get back to this, because this is an important thing in and of itself.

I think this notion of prions or mycoplasma being a thing might have a a bit of truth to it.

Even if you don't totally accept the definition of viruses as handed us by germ theory, this seems to be something concrete that there might be something to.

1:05:43

We'll see.

I could be wrong.

I do reserve the right to be totally wrong about that, but I just find the connections here very interesting.

So it says here is.

It is possible that the crystalline disease toxin from the pathogen is one of the Mycoplasma species, a technological feat accomplished by U.S. military biochemical researchers working with Nazi paper clip scientists.

1:06:06

In 1946, the director of the War Research Service, George W Merck, reported the possibility of using crystalline toxins to Secretary of War Robert P Patterson.

It should be noted that the War Research Service initiated America's biological weapons program and Merck went on to become president of the Merck and Company pharmaceutical firm.

1:06:30

Going to pause for a moment here, folks.

Did you know that?

Did you know that Merck Pharmaceutical found its origins developing biological warfare weapons?

And they produce a lot of your medicines, quote, UN quote medicines today?

1:06:50

Anyway, let's continue on.

Although Merck died in 1957, his early knowledge of the disease toxin means it could have been passed along to his colleagues at Merck Pharmaceutical.

That Merck was involved in such research can be seen in a New England Journal of Medicine article that noted that a study of the hepatitis B vaccine used extensively in gay and drug addict communities was supported by a grant from the Department of Virus and Cell Biology of Merck, Sharp, And Dome Research Laboratories, West Point, VA.

1:07:26

Going to pause for a moment here folks.

So the Department of Virus and Cell Biology of Merck, Sharp, and Dome Research Laboratories, they offered a grant to develop the hepatitis B vaccine.

1:07:44

Interesting, right?

Let's continue on.

After extensive study, researchers Donald West and William LC Scott concluded that those suffering from chronic fatigue syndrome and fibromyalgia are actually victims of, quote man, altered versions of brucellosis emanating from the triangle.

1:08:07

That is the areas found around Fort Detrick, Washington, DC, New York City's East Side and Long Island's Federal Animal Disease Center and Cold Spring Harbor Laboratory.

End Quote.

These locations are often mentioned in biological warfare literature.

1:08:25

Fort Detrick and Cold Spring Harbor especially were centers of Nazi paper clip research activity.

Going to pause for a moment here, folks.

And there's also a little place called Plum Island near Long Island that is the alleged home of the birth of Lyme's disease.

1:08:50

Keep that in mind.

Let's continue on here.

According to the Scott's report, this pathogen was tested during the summer of 1984 at Tahoe Truckee High School in California via the air duct system.

1:09:08

Gonna pause for a moment here, folks.

So were you aware of this?

Did you know that they were doing this type of testing without the consent or knowledge of the people as recently as 1984?

1:09:28

You see, you mentioned something like that to people.

And they think, oh, yeah, well, maybe they were doing the thing with the syphilis back, you know, in the World War 2 era or something.

But they've not.

They wouldn't do anything like that today.

Yeah, you want to bet?

I would think people are largely unaware of this stuff.

1:09:47

But these are admitted things now, so Tahoe Truckee High School in California.

In their air duct system, they tested this pathogen.

Let's read on says Individual rooms were fitted with an independent recycling air supply system, and the teachers lounge was designated as the infection target.

1:10:11

Within months, seven of eight teachers assigned to this room became very ill Tahoe Truckee High School was only one of several locations where these specially designed pathogens were tested.

Some pathogens were distributed by aerosol sprays and others were spread through contaminated mosquitoes.

1:10:31

Gonna pause for a moment here, folks.

There's nothing new under the sun.

Bill Gates is not the first guy to genetically engineer mosquitoes and set them loose on the world, or to put attack vectors in mosquitoes and set them loose.

The old tricks are the best tricks, folks.

1:10:48

They've been doing this stuff for a very long time, very long time.

And of course I'll think of a more recent event several years back.

Now forget exactly when.

Do you remember they tested some gas, some strange gas in the New York subway system, and they said, of course it was harmless.

1:11:11

Was it?

Was it harmless, do you think?

And this was a visible test that they did with this.

I forget what the the whole notion, the excuse for doing it was, but they they claim it was totally safe.

1:11:27

It was nothing.

It was nothing to be worried about.

And that they sprayed this gas in the subway system in New York City.

Anyway, you could look that one up yourselves too, but let's continue on because we're running low on time here.

So the Scots reported that during the 1980s, one 100 million mosquitoes a month were bred at the Dominion Parasite Laboratory in Bellevue, Ontario.

1:11:53

From there, the mosquitoes were tested by both Canadian and U.S. military authorities after being infected with brucellosis.

Some observers believe the 1999 outbreak of human encephalitis in New York City, due to what was designated, quote, UN quote, West Nile virus, may have been the result of these infected mosquitoes.

1:12:14

Going to pause for a moment here, folks.

Once again, we're seeing all these connections that seem to keep cropping up here.

Are they truly identifying different diseases and illnesses in certain ways and treating them with certain medications or pharmaceuticals when it's really just all this one thing this man made biological weapon developed by the Nazi project paper clip scientists and unleashed on the public in experiments.

1:12:55

Could it be true?

I don't claim to know the answers to that, but like I said, comes down to asking the right questions.

First of all, do you trust your government?

Do you trust Big Pharma?

Do you trust the multinational corporations that are making a killing in profits on your medical treatment?

1:13:16

And do you trust these nonprofit organizations that seem to steer in direct policy across the world, even if you give them the shadow of the doubt?

And you want to maybe be optimistic and pie in the sky and think, well, they're just good people trying to do the right thing.

1:13:37

They keep proving you wrong over and over and over again, disappointing and proving you wrong on that notion.

They lie, they cheat, they steal, they manipulate.

And yes, they have murdered folks in the past to keep their secrets, no doubt.

1:14:01

So do you put something like this past these people, these select few at the top of the controlling power structure here?

Do you really think they would not do something like this?

They would never do that.

Well no, you would never do that.

1:14:18

They're psychopaths.

They most certainly would.

They believe in eugenics, they act on eugenics.

They actively avow.

They want to see the population reduced by, oh, about 7 billion people or so.

And how do they want to do it?

1:14:38

Well, through vaccines and medical care.

Bill Gates in a Ted Talk talking about reducing that number by 15% with good vaccines and medical care.

Yeah.

OK And these people want to help you.

1:14:54

Really.

I don't think so.

Let's read on here because we're running low on time and there's still a bit more ground I'd like to cover here before we sign off.

Additionally, the Scots also claim that unsuspecting victims were tested by both the military and CIA and monitored by the National Institutes of Health in the Centers for Disease Control.

1:15:15

Encouraged by what they thought was a successful test, military leaders reportedly passed the Bruce Cellosis bio agent to Saddam Hussein, who in the mid 1980s was fighting a protracted war against Iran with the aid of the CIA.

With the approval of Vice President George HW Bush in 1985, Saddam received, quote, a startling array of biological pathogens, the essential raw material for a disabling weapon.

1:15:43

End Quote.

This included shipments of both Brucella Abortis biotypes 3 and 9 and Brucella Tensis biotypes one and three.

These toxins continued to be sold to Saddam through May 2nd, 1986 as quote shipments #21 and 22 from the American Type Culture Collection ATCC in Rockville, MD.

1:16:11

End Quote.

Notice the number 22 prominent there.

They put their fingerprint upon this, their stamp of approval, those dark occultists who run things.

And it was America that supplied the bio weapons to Saddam.

1:16:32

In a 2005 article entitled Molecular Terrorism, Gary Tunsky credited both the Scots and the Nicholsons with creating a growing public awareness of the mysterious and debilitating effects of microplasma infection.

Chances are if you feel sick and tired and your doctor is unable to make a definite diagnosis because lab tests, blood chemistry profiles and tissue cultures failed to reveal any disease pathogen, you might very well be infected with Mycoplasma, suggests Tonski.

1:17:03

Since Mycoplasma cannot be successfully treated with the usual sort of short course duration of antibiotics due to their intracellular location, slow proliferation rate and inherent resistance to most antibiotics, The few Mycoplasma experts that specialize in this field are recommending six months to one year of non-stop treatments using strong antibiotics such as Cipro and doxycycline, he added.

1:17:29

However, if a patient does not want to destroy their body and immune system with Cipro and doxycycline, a total overhaul of every cell from head to toe using a multifaceted, non-toxic, holistic treatment approach is absolutely necessary to overcome Mycoplasma infections naturally.

1:17:49

This is why vitamins and nutritional supplementation are so important in the therapy, Tonski said.

The reason so many Americans are caught up in a medical merry go round of being bounced from one doctor to the next without ever receiving a proper diagnosis is the mainstream medical doctors are not trained to find hard to detect pathogens.

1:18:10

Since Mycoplasma hides intracellularly and invades multiple organs and systems, it manifests a very vast array of symptoms throughout the whole body, making incorrect diagnostics virtually impossible for a mainstream doctor's linear magic bullet mentality, he explained.

1:18:30

Gonna pause for a second.

Such inability to make a quick and simple diagnosis lies behind the mysterious malady that struck members of the US military in the Persian Gulf War of 1990 to 1991.

And of course, folks, that was called Gulf War Syndrome, which of course may be traceable back to this brucellosis, this weaponized version of brucellosis, these mycoplasmas, these prions as they're called in their various forms.

1:19:09

Maybe that is the thing that underlies all of these many things.

You really have to wonder.

This stuff obviously has been in development and experimentation since 1946, bare minimum, probably before that, because I I guess the Nazi scientists developed it towards the end of the war and they brought all their research here to the US in 1946.

1:19:39

And since then we see the rise in unwellness across all of Western culture, especially here in America.

And I think there's a lot of different synergistic operating principles behind that.

1:19:58

I think it's a combination of, first of all, the crap they're spraying us with in the sky, the foods that we're eating, all the substances that they're adding to our foods.

And by the way, they're really pushing heavily and they're beginning to add insect proteins to our foods.

1:20:15

Now remember one of the original attack vectors that they used, this crystalline form of the brucellosis.

In these prions in was insects.

So they're putting insects in our food without our consent and without our agreeing to it, without our knowledge.

1:20:42

Likely there might be something to that too.

So we have these different combined factors that seemed to contribute to our own Wellness.

We have the crap they're spraying us with in the sky, all the garbage they're putting in our food and water.

We have the frequency bands that we're being inundated with.

1:21:04

We're swimming in a soup of microwave radiation of all different sorts right now, folks, even if you don't have Wi-Fi in your house or any such thing, the frequency bands that we're swimming in are just unbelievable.

There's so much of it around.

1:21:21

So that being the case, we have all these different vector points that may be triggering mechanisms for such a thing occurring.

So that might be the case.

As to what's going on here, is there something to this connection?

1:21:38

To prions?

To mycoplasma?

To this brucellosis?

This weaponized version of brucellosis developed by the Nazi scientists and continued by the paper clip scientists?

Maybe.

I think it's definitely worth investigation.

Will you ever see an official investigation of this happen?

1:21:56

Probably not.

Will your doctor even have a clue what you're talking about if you tell him about this?

Probably not.

Are there treatment options available?

That's the other question.

How would you treat something like this aside from heavy duty, long term use antibiotics as presented here.

1:22:18

I don't know of any naturopathic ways to get rid of this.

Not saying that there's not a way to do so.

Perhaps somebody out there knows more about that than myself.

Maybe there is and maybe that's the way to go.

1:22:35

I would think anything that you would do to combat this would probably be some kind of a long term treatment plan.

It's not going to be gone overnight because of the nature of it.

It's stealth.

It hides in your cells, can't be detected by any test.

1:22:55

Unless, of course, maybe you could find it through the use of the PCR test, or maybe you could make sure it's there through the use of the PCR test.

Maybe you know there's something to that.

But there's all these different things that make us unwell, and a lot of them could be triggering mechanisms for some such stealth bio weapon to be activated in people.

1:23:25

And who knows, maybe they were directly injecting people with this bio weapon.

Can't say for sure, Don't wanna go making accusations, but but it is worth consideration and it is worth investigation.

1:23:44

Especially if you have all of these sources that can tell you these things and document these things.

Project Paper clip was most certainly a real thing.

We did have Nazi scientists come over here and they continued their research and they performed all kinds of heinous experiments.

1:24:06

And we have documentation of the American government performing experiments on the people without their knowledge or consent as well.

We have that.

We have these records.

I mean, this is historical record.

This is not disputed.

This is mainstream history.

1:24:22

It's a known commodity.

It's indisputable even by their own standards.

And yet, are we going to still trust these people?

At the end of the day, it all falls down to that big question again.

Do you trust your government?

Do you trust Big Pharma?

1:24:41

Do you trust multinational corporations that have a vested interest in your not being?

Well, do you have trust for those nonprofit organizations that seemingly run our world and set up our foreign policy and our domestic policy and seem to do all kinds of things without our permission or consent speaking on our behalf?

1:25:05

Do you really trust them?

Have they shown us that they are trustworthy, that they're on the level to use their terminology?

Are they really?

I think the more we look, the more corruption we find, and the more we find they are not trustworthy.

1:25:25

The media machine that they own tries to cover for them and does an indispensable job for them of doing so, of spin doctoring and of trying to save face for them and deflect blame everywhere.

1:25:45

So we have to wonder, is there something to this notion?

I think maybe there's a kernel of truth involved.

Can't say for sure.

It would require an investigative strategy that I myself am unable to undergo.

1:26:01

I don't have the resources to do so, and therefore it's probably something we'll never get a proper answer to.

It's probably something that will never be properly addressed by anyone, and the status quo will continue.

1:26:21

The thing is, I think people by and large here in the world are sick and tired of the medical mafia and are sick and tired of the whole pandemic narrative, and they're not afraid anymore.

1:26:37

And that's problematic for these people.

So even if they are waging a type of biological warfare campaign against the public, people aren't afraid of it anymore and they're not going to capitulate to some of the demands.

1:26:52

So of course they're shifting gears and of course they've always got to beat the war drum.

It's always the time tested method for social change, war, so two giant proxy wars going on.

And of course push the fear notion across the nation here, associated with that to create instability, to muster up more fear in the minds of the people.

1:27:25

Make them skeptical about leaving their homes, make them lock themselves in their homes, lock down their businesses willfully for whatever reason, for fear of harm of some sort.

It's just a different, different type now.

1:27:41

They're not going to use the same threat two times in a row, seems to me.

They're throwing everything against the wall and seeing what sticks at this point.

So that's what we're getting.

And of course, war is always the time tested way to get things done for these people.

1:27:57

Oftentimes it's their last resort for implementing social change the way they want, and we're seeing that.

So just be mindful.

Get your heart right with God, folks.

Get your relationship right with God.

We're living in interesting times for sure.

1:28:15

And keep your eye on what's going on out there and understand everything that comes across your television screen is a manipulation.

And when you look at it from that approach, you understand that you're not being given the true nature of what's going on.

1:28:32

It's all through the lens.

They want you to see it.

And that's where we're at in this reality that we live in.

They feed us what they want us to see and it's all about do we turn off the program and think for ourselves or do we just go along with the 2 little boxes they want to throw us in?

1:28:53

You see they always have two sides.

You pick a side, They want you to pick a side with this stuff and go there and then they could steer and manipulate you in whatever direction that way.

Don't think for yourself, just go along to get along, be agreeable.

1:29:11

That's what they want anyway.

Folks, I want to thank you all for tuning in tonight.

I appreciate each and everyone of you.

We'll catch you next time.

Have a good night.

Visit www.alchemicaltechrevolution.com to further support my work. Thank you and God bless you all.

www.ingramcontent.com/pod-product-compliance
Lightning Source LLC
Chambersburg PA
CBHW062236290526
45794CB00006B/2303

* 9 7 9 8 8 6 6 2 9 6 9 6 5 *